LEARN TO COUNT & WRITE NUMBERS WITH ME

We create our books with love and great care.

Yet mistakes can always happen. For any issues with your book, such as faulty binding, printing errors, or something else, please do not hesitate to contact us at: **hello@eduplayclub.com**
We will make sure you get a replacement copy immediately.

For any suggestions or questions regarding our books, please contact us at: **hello@eduplayclub.com**

ISBN: 978-8412468366

Without your voice we don't exist.
Please, support us and leave a review!

Thank you!

This book belongs to:

Welcome to Edu Play Club!

The club for caring parents and happy children.

At Edu Play Club we believe that learning can and should be fun. This is our philosophy when we create each of our educational books, aiming to make them wonderfully fun and engaging to help you raise kids who love to learn.

Write to us at: **hello@eduplayclub.com**

Tell us what you think about this book. We listen. We take every suggestion into account and try to implement it whenever it is possible. Thank you for helping us make our books better!

EDUPLAYCLUB

Follow us on **Instagram:**
@eduplayclub

We will share regularly information, recommendations, useful resources, inspiring ideas, learning activities, free printables and of course news about our upcoming books.

A word for the parents:

Every child's learning journey is unique and unfolds at its own pace. Don't push your child into activities they're not ready for.

Celebrate each effort, whether it's correctly counting, tracing a number for the first time, or simply showing curiosity about the activities. Encouraging your child's efforts rather than expecting perfection helps foster a love of learning and builds confidence.

Thank you

for being a part of our journey towards creating fun, inspiring, and enriching books for children everywhere!

Please, take just a moment of your time to leave a review. It makes a big difference and helps us immensely.

Review here

1

2

3

4

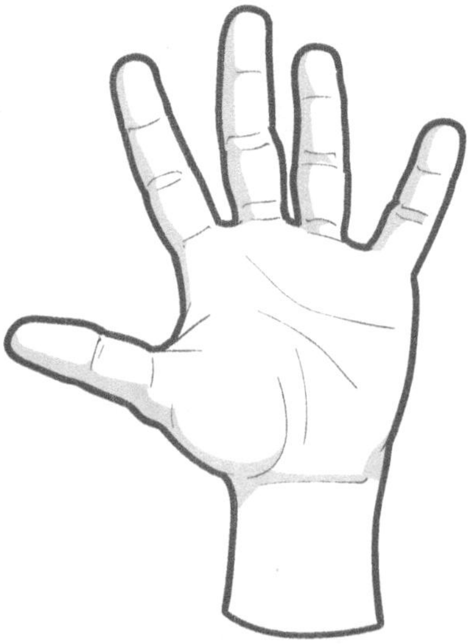

5

Count and color!

1

one

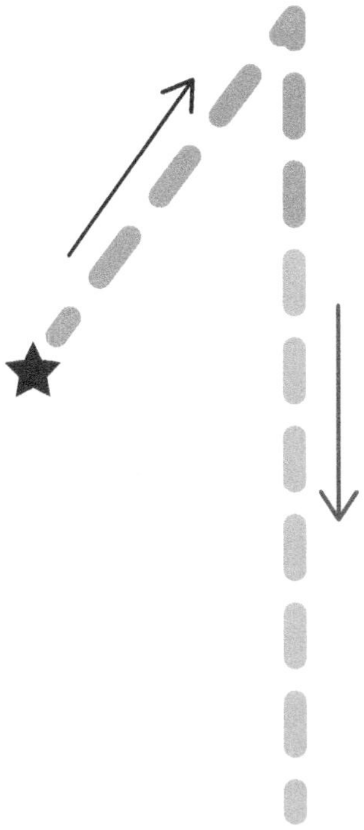

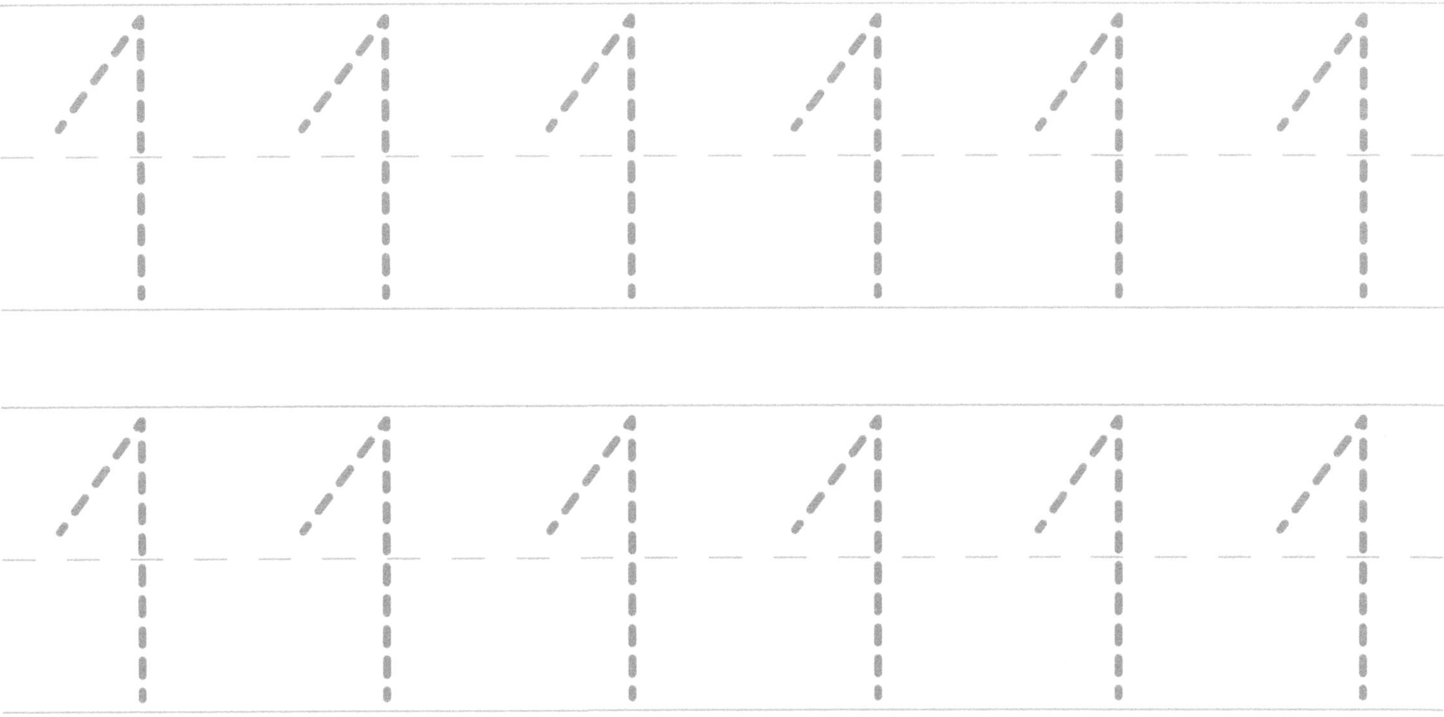

1 1 1 1 1 1

1 1 1 1 1 1

1 1 1 1 1 1

1 1 1 1 1 1

1 1 1 1 1 1

There are items that appear only 1 time in the picture. Color them.

2

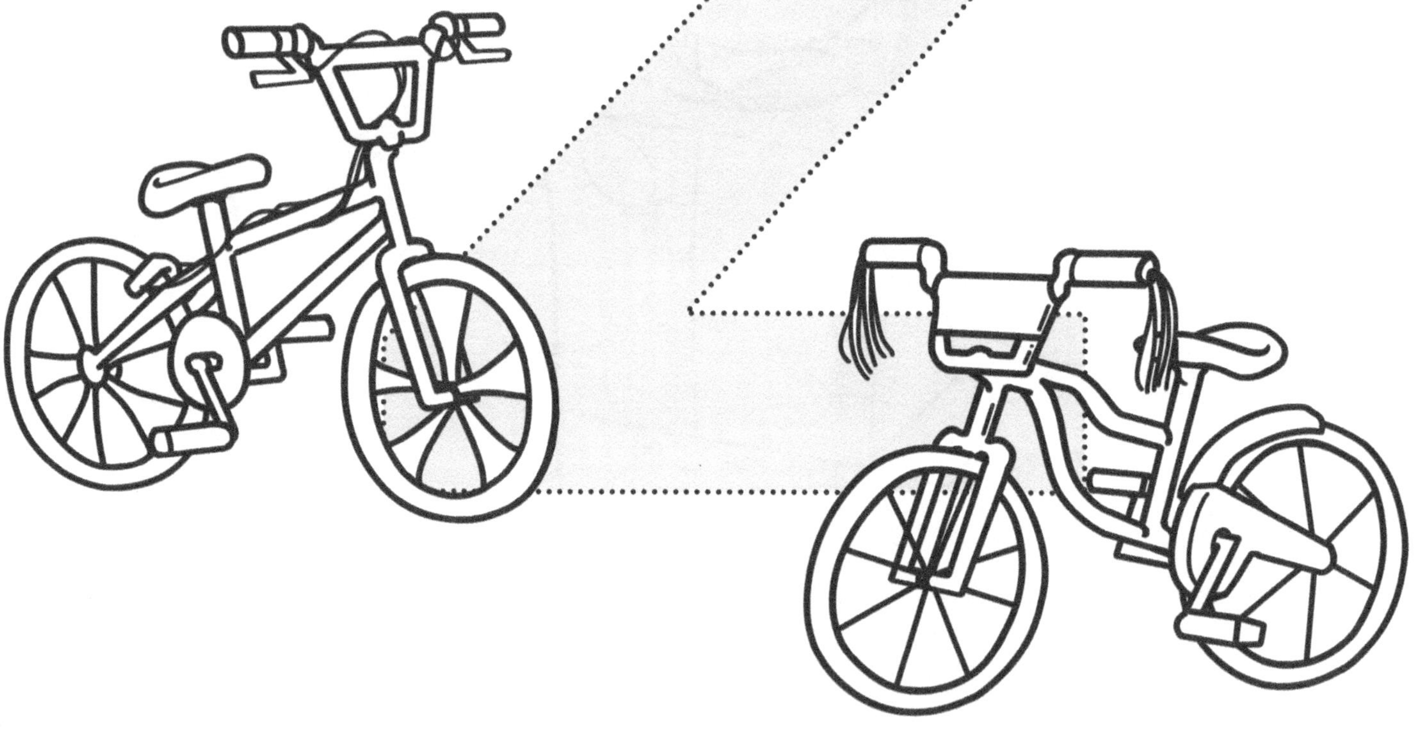

two

2 2 2 2 2 2

2 2 2 2 2 2

Draw the missing 2 wheels on the bicycle and color the picture.

three

3 3 3 3 3 3

3 3 3 3 3 3

3 3 3 3 3 3

3 3 3 3 3 3

3 3 3 3 3 3

Trace and color the 3 butterflies.

four

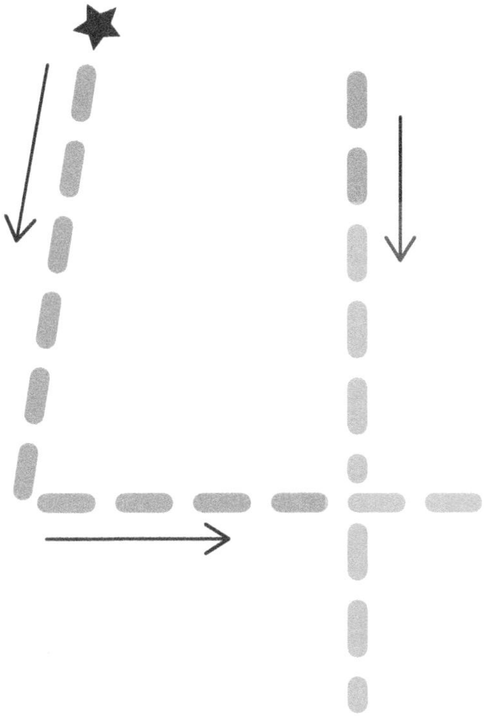

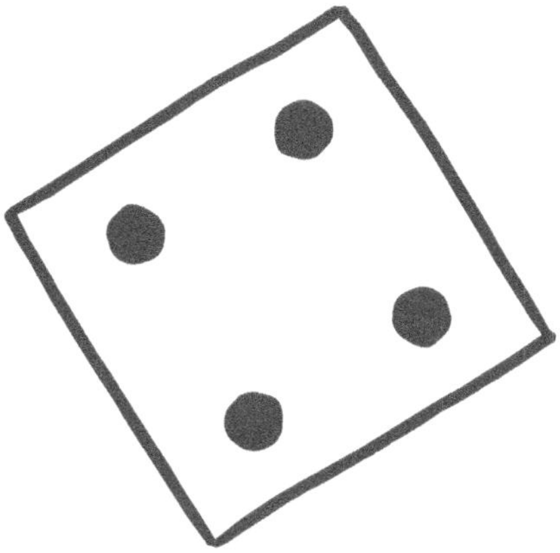

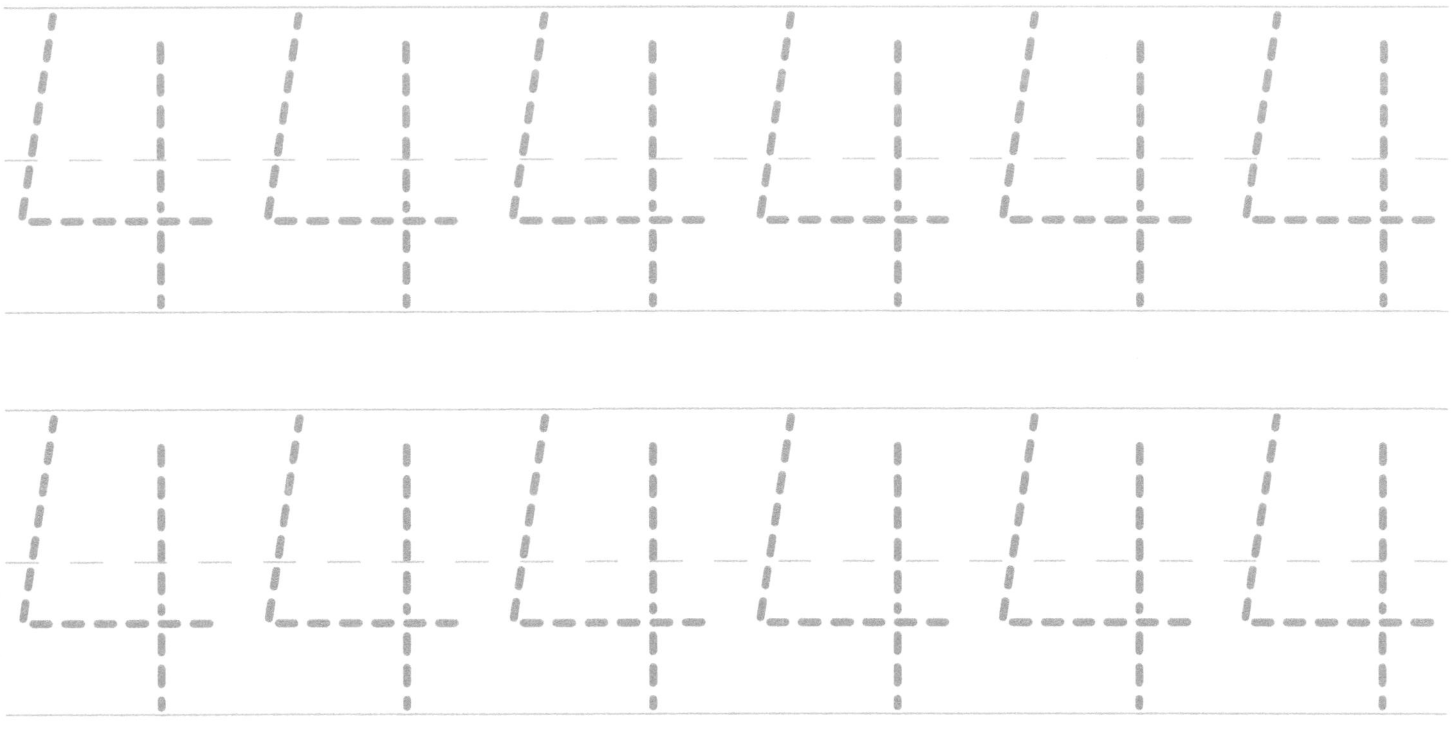

4 4 4 4 4 4

4 4 4 4 4 4

4 4 4 4 4 4

4 4 4 4 4 4

4 4 4 4 4 4

Which animals have 4 legs?
Color them.

five

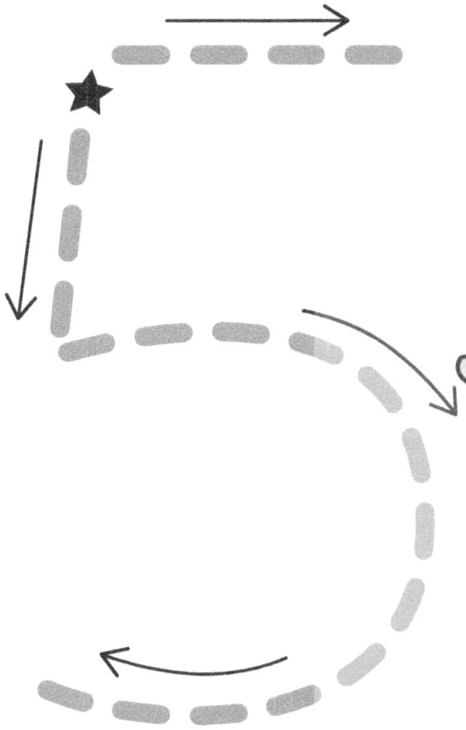

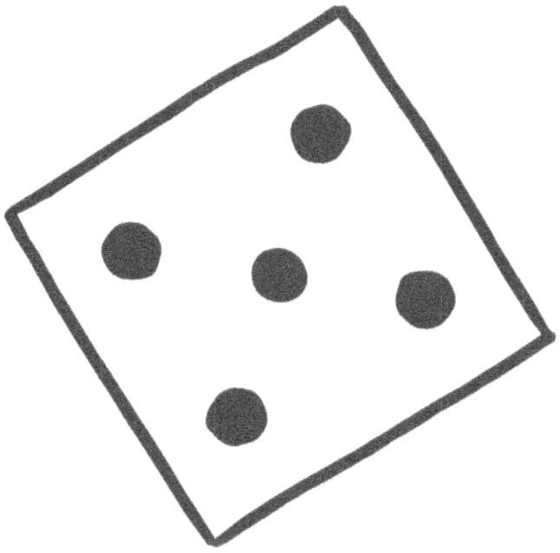

5 5 5 5 5 5

5 5 5 5 5 5

5 5 5 5 5 5

5 5 5 5 5 5

5 5 5 5 5 5

5 5 5 5 5 5

5 5 5 5 5 5

Trace the 5 Olympic rings.
Also, trace the Paralympic symbol
and color it in red, blue, and green.

Count and match!

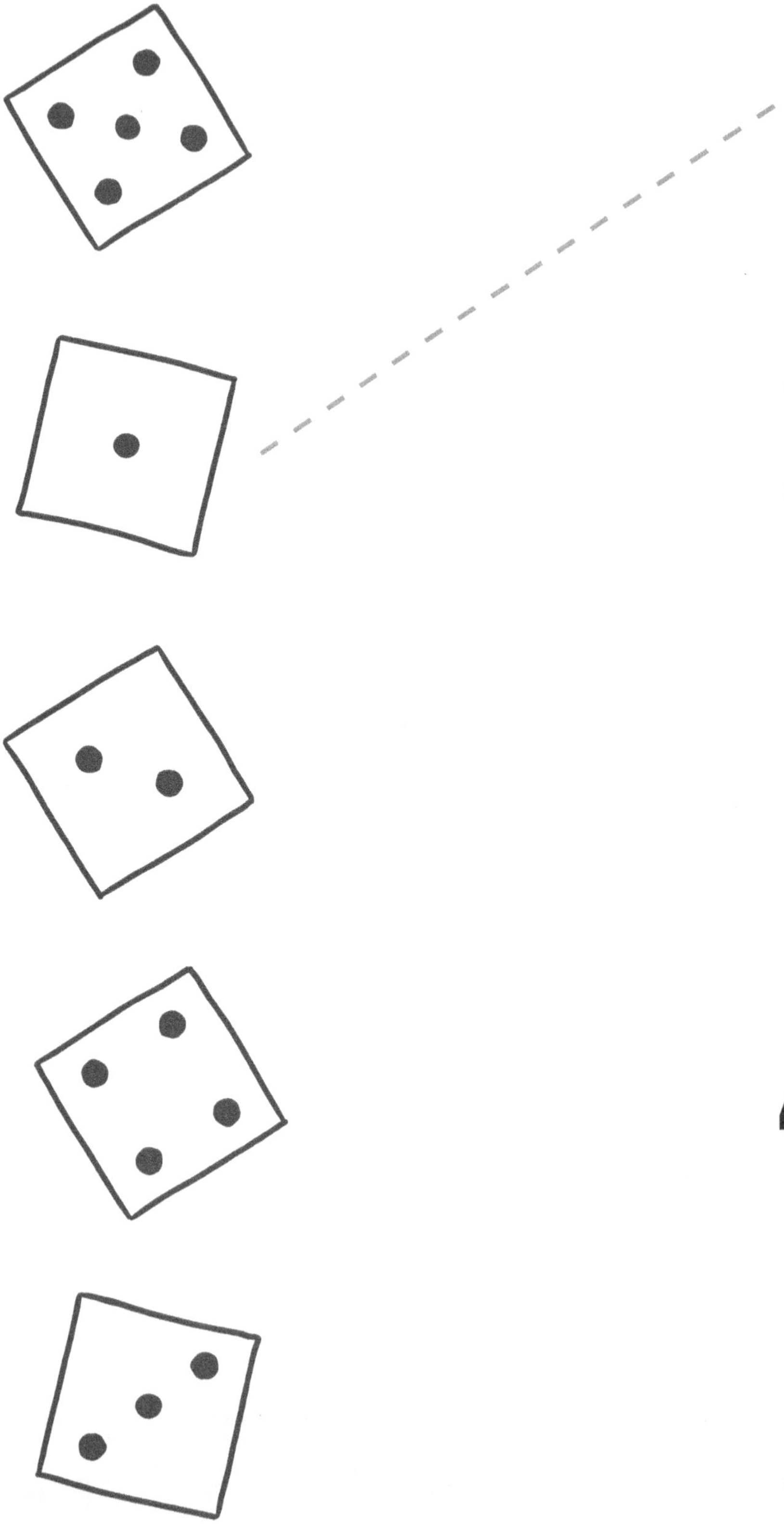

1

2

3

4

5

Count and match!

1

2

3

4

5

Count and write!

2

Color by number!

green	1
purple	2
yellow	3
red	4
orange	5

6 

7 

8 

9 

10 

Count and color!

six

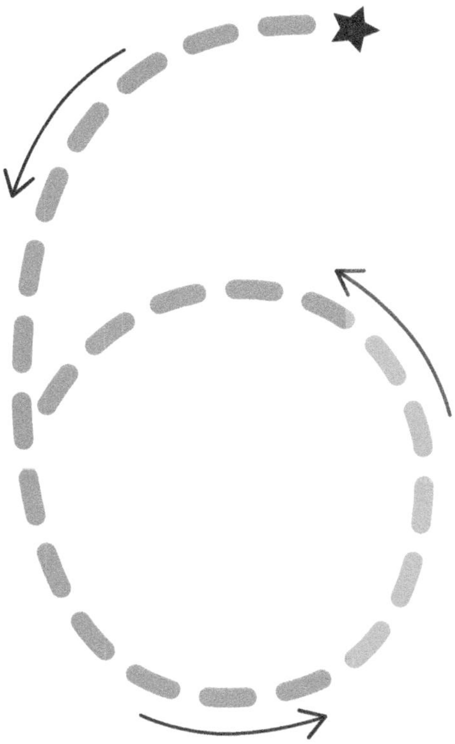

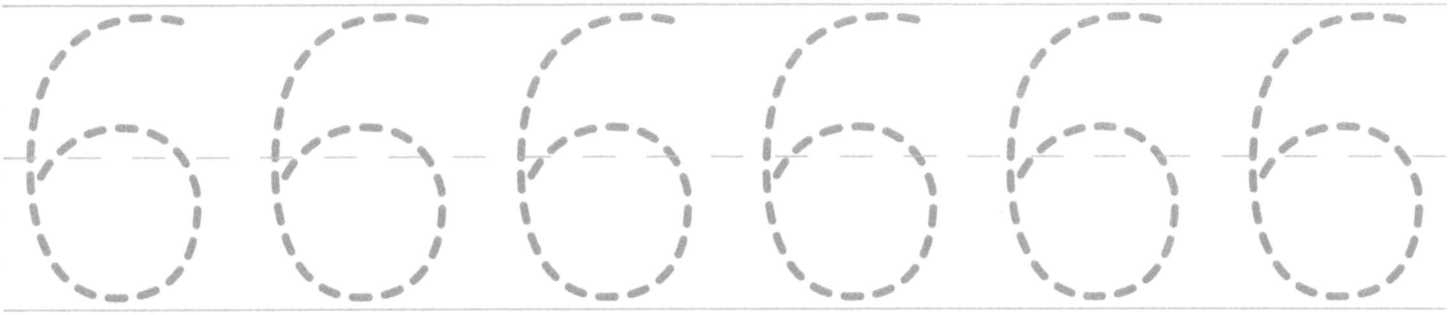

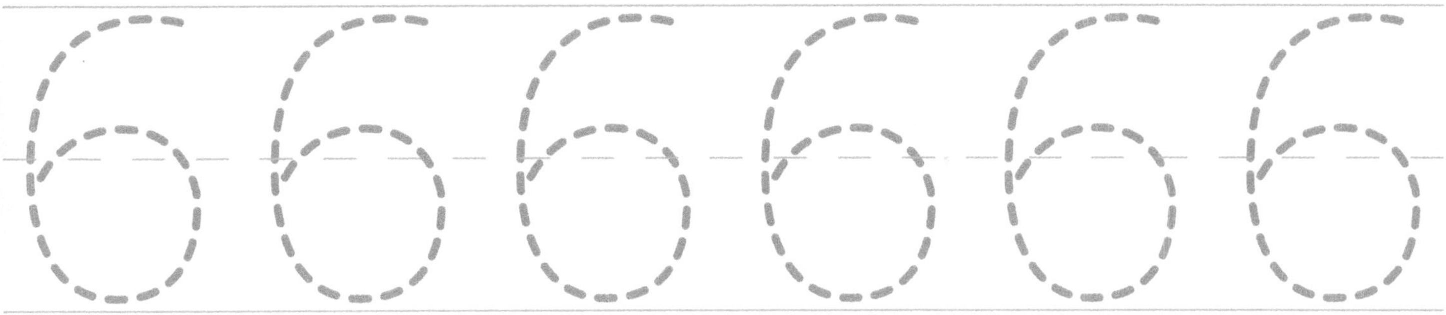

6 6 6 6 6 6

6 6 6 6 6 6

6 6 6 6 6 6

6 6 6 6 6 6

6 6 6 6 6 6

Color all 6 items that the boy needs for his knights and dragons game.

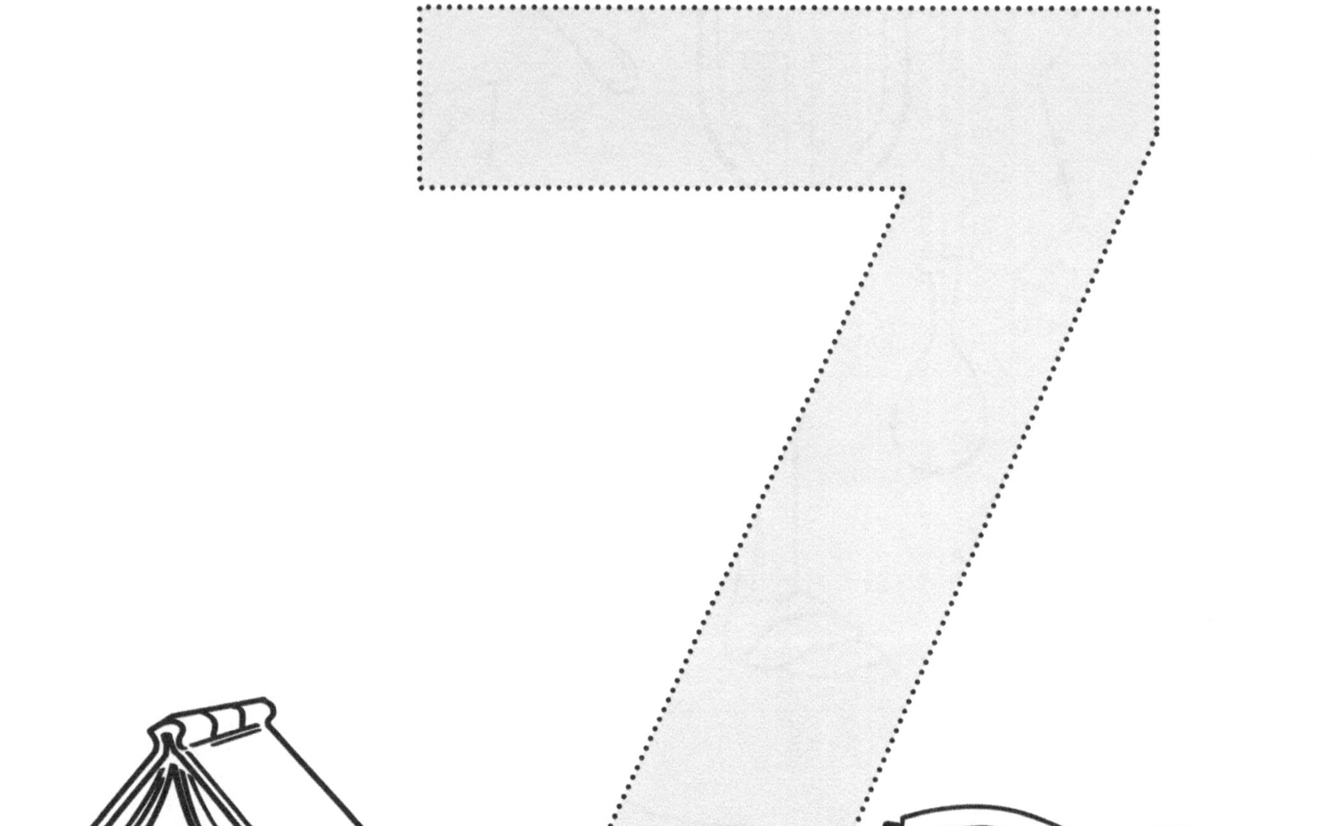

seven

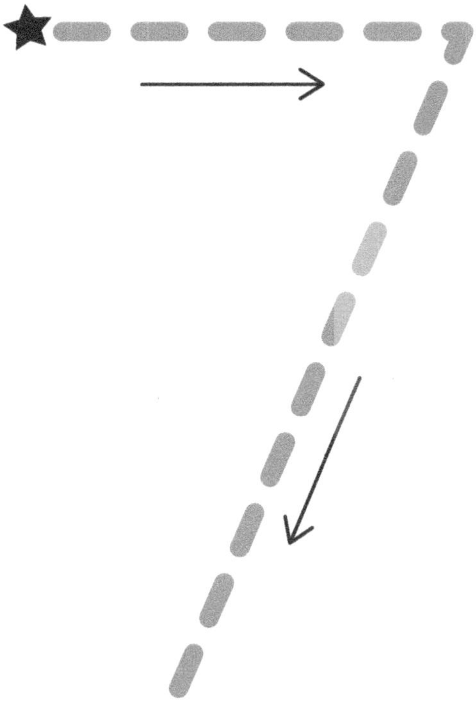

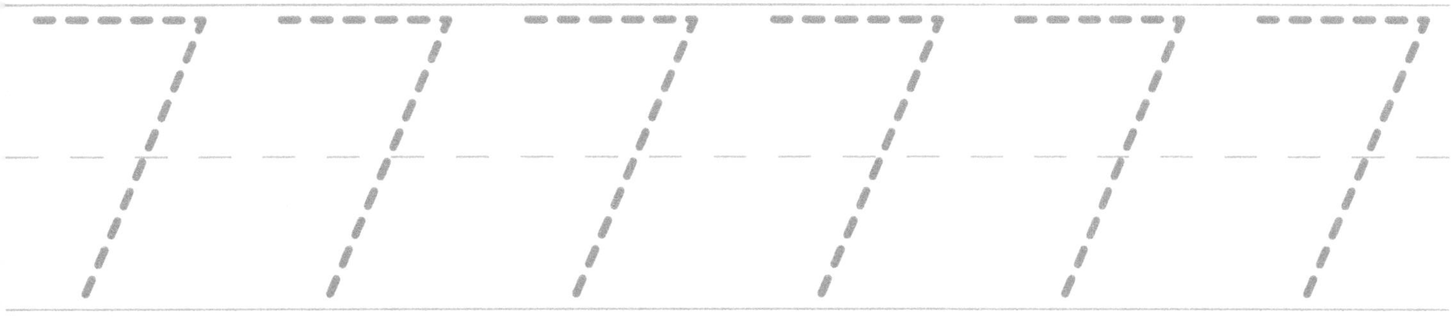

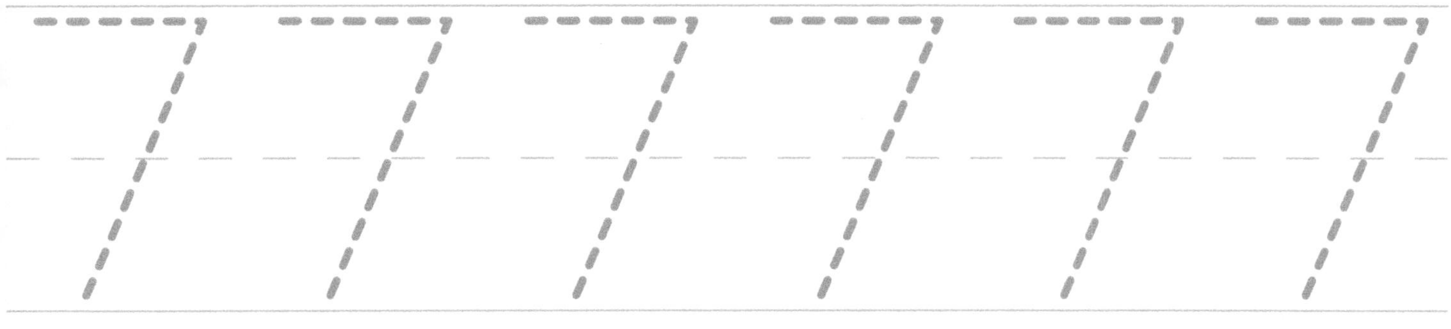

7 7 7 7 7 7

7 7 7 7 7 7

7 7 7 7 7 7

7 7 7 7 7 7

7 7 7 7 7 7

Find and color all 7 parts required to build the robot.

eight

8 8 8 8 8 8

8 8 8 8 8 8

8 8 8 8 8 8

8 8 8 8 8 8

8 8 8 8 8 8

Find and color the 8 fish that are the same.

nine

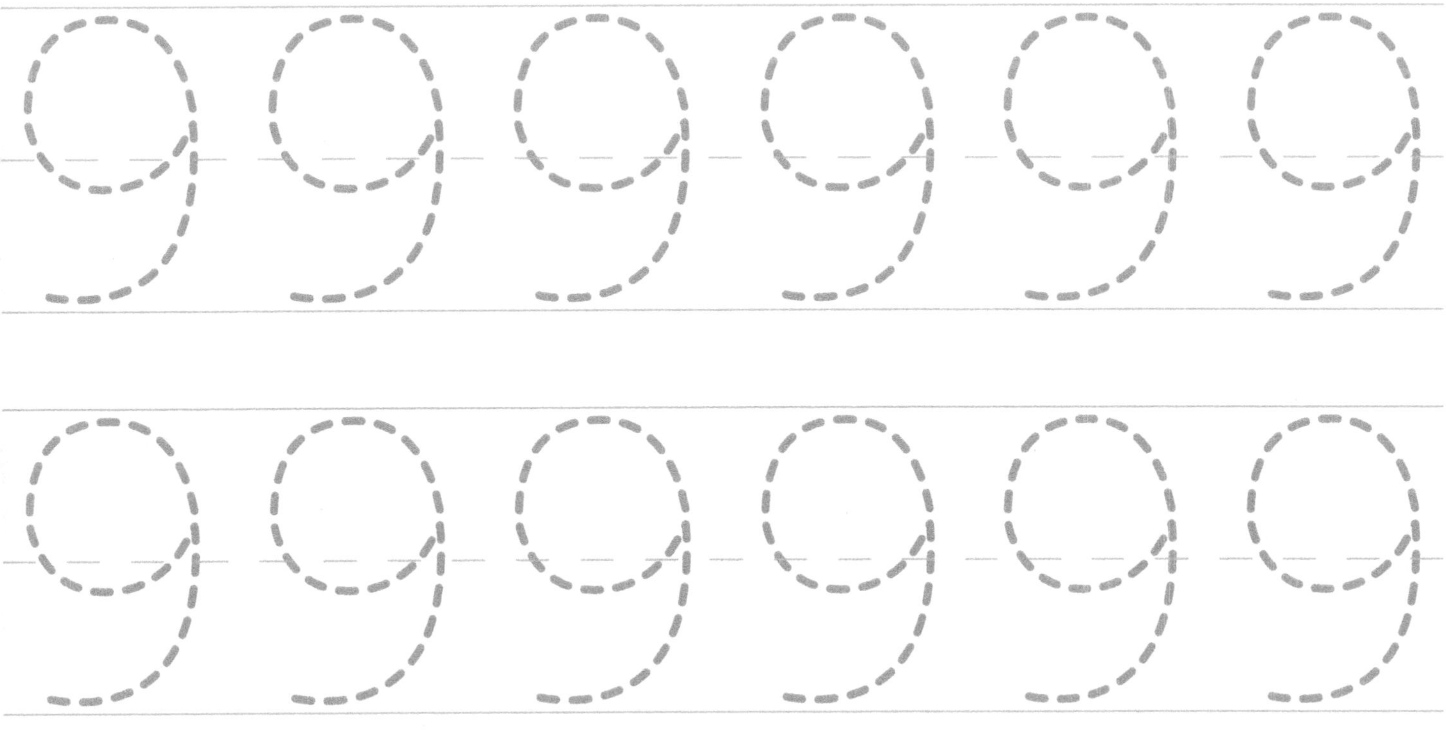

9 9 9 9 9 9

9 9 9 9 9 9

9 9 9 9 9 9

9 9 9 9 9 9

9 9 9 9 9 9

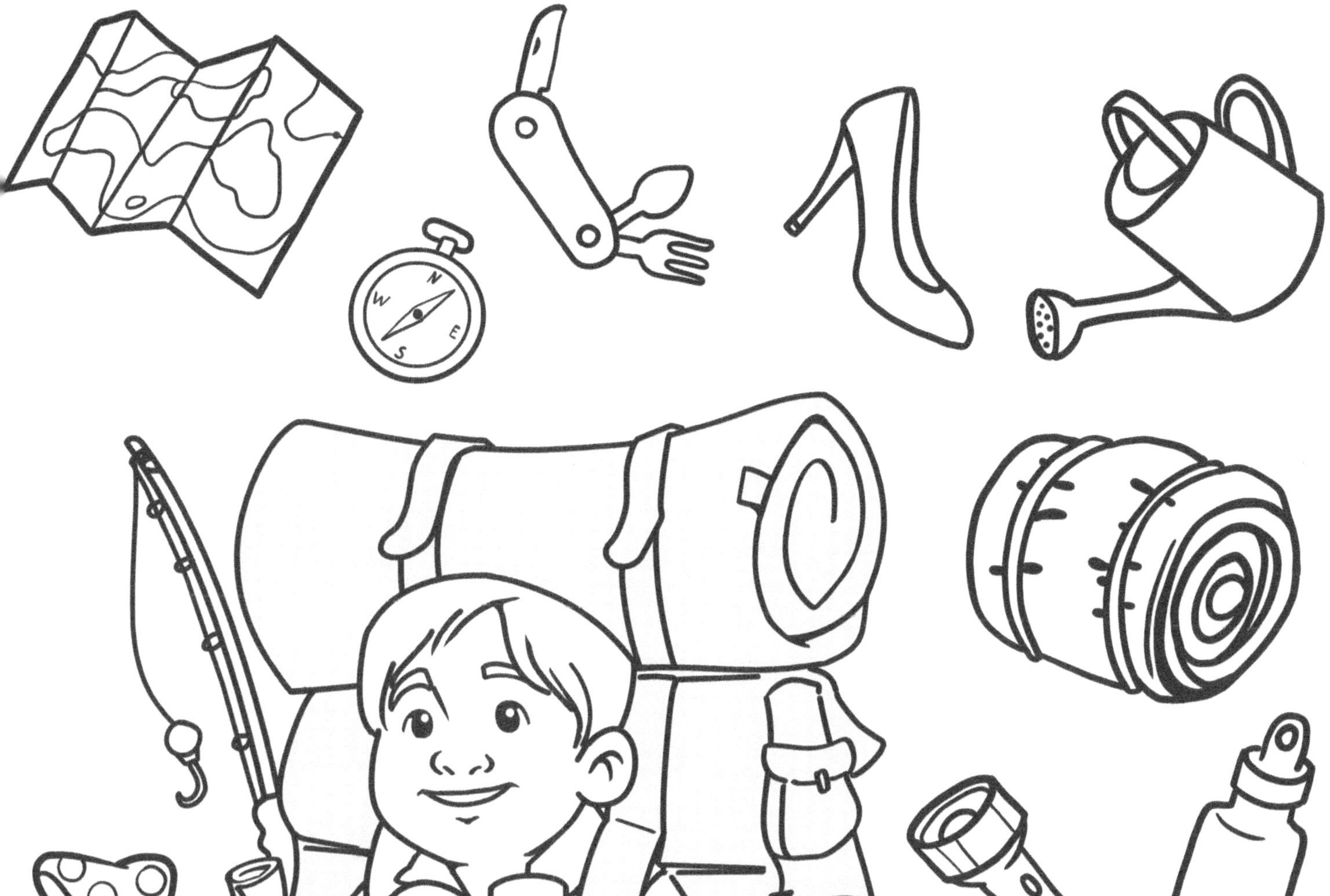

Choose 9 objects to put in the backpack and color them.

ten

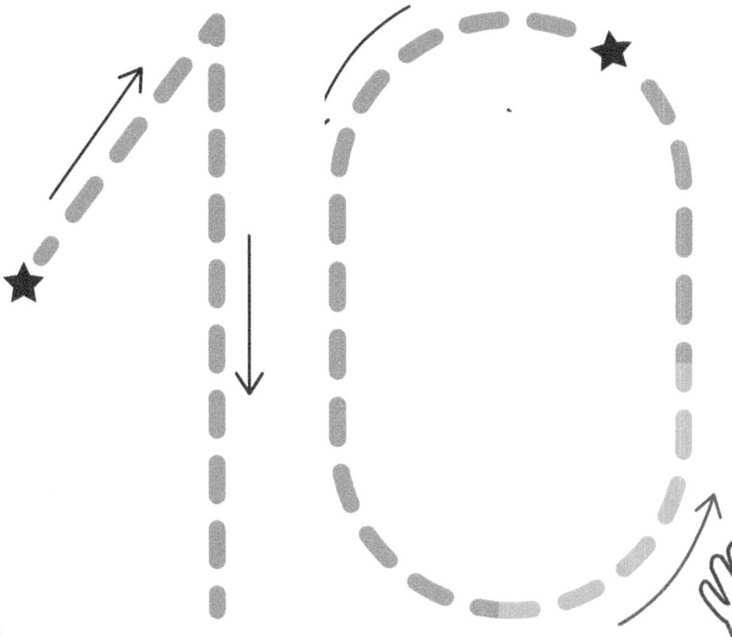

10 10 10 10

10 10 10 10

10 10 10 10

10 10 10 10

10 10 10 10

10 10 10 10

Draw more eggs. The girl needs 10 to make cookies.

Count and match!

6

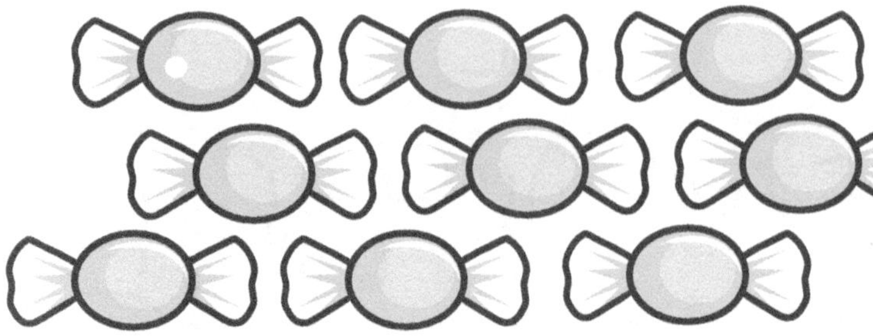

7

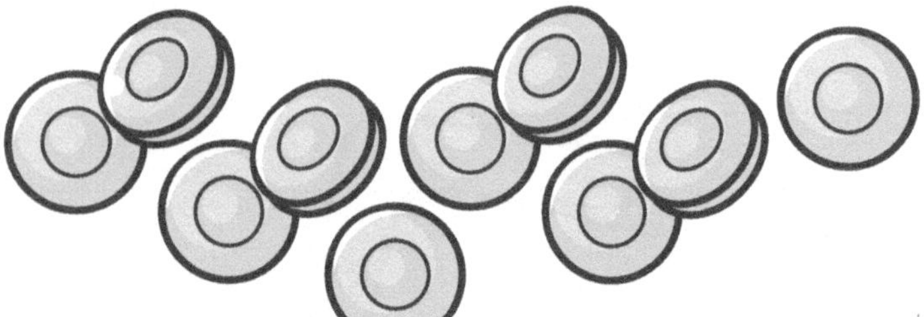

8

9

10

Count and write!

1
5

Fill in the missing numbers.

eleven

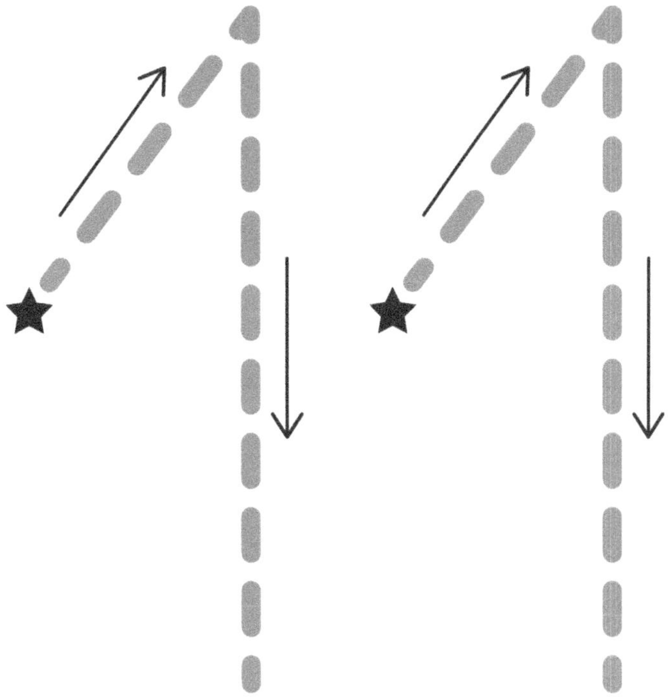

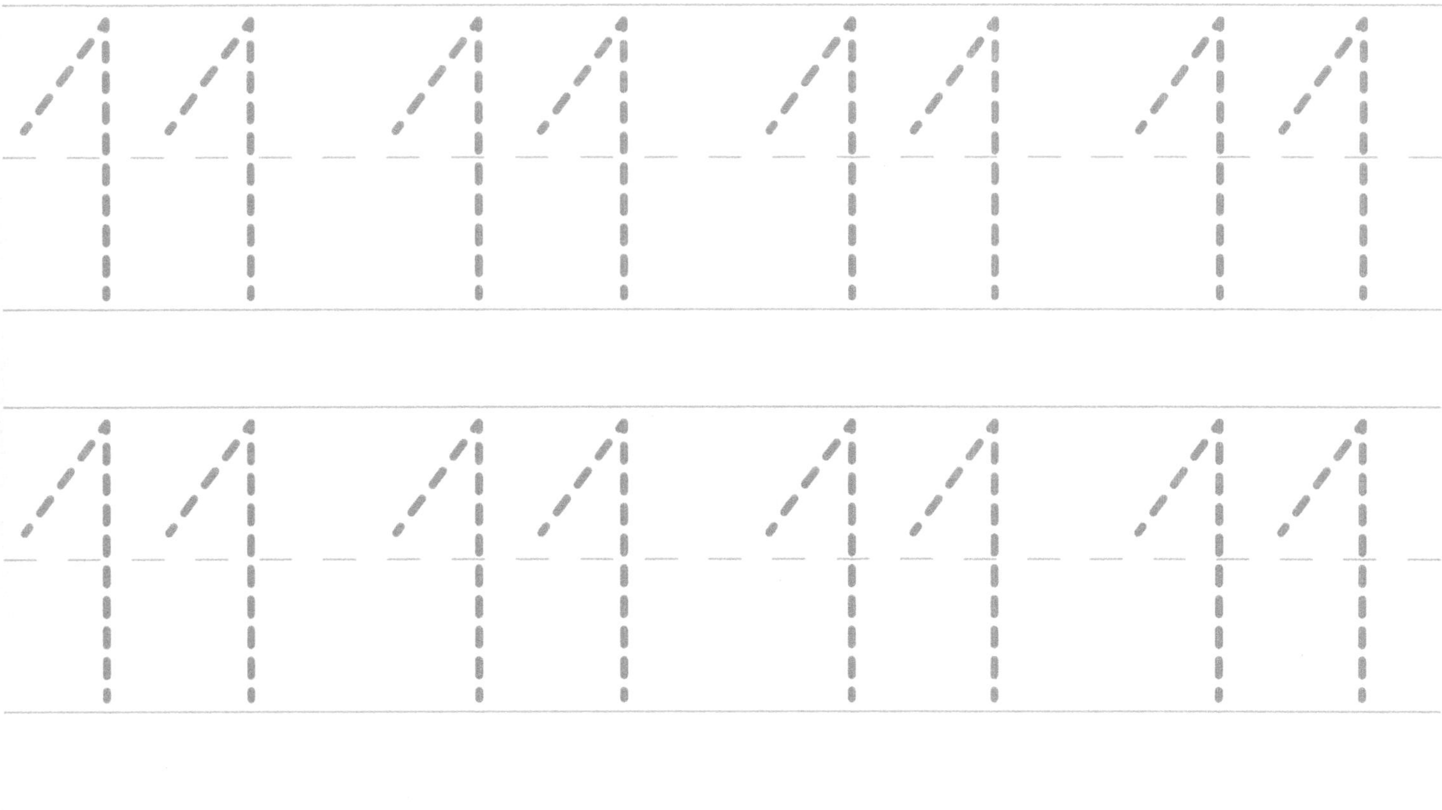

11 11 11 11

11 11 11 11

11 11 11 11

11 11 11 11

Trace the 11 paper planes and color them.

twelve

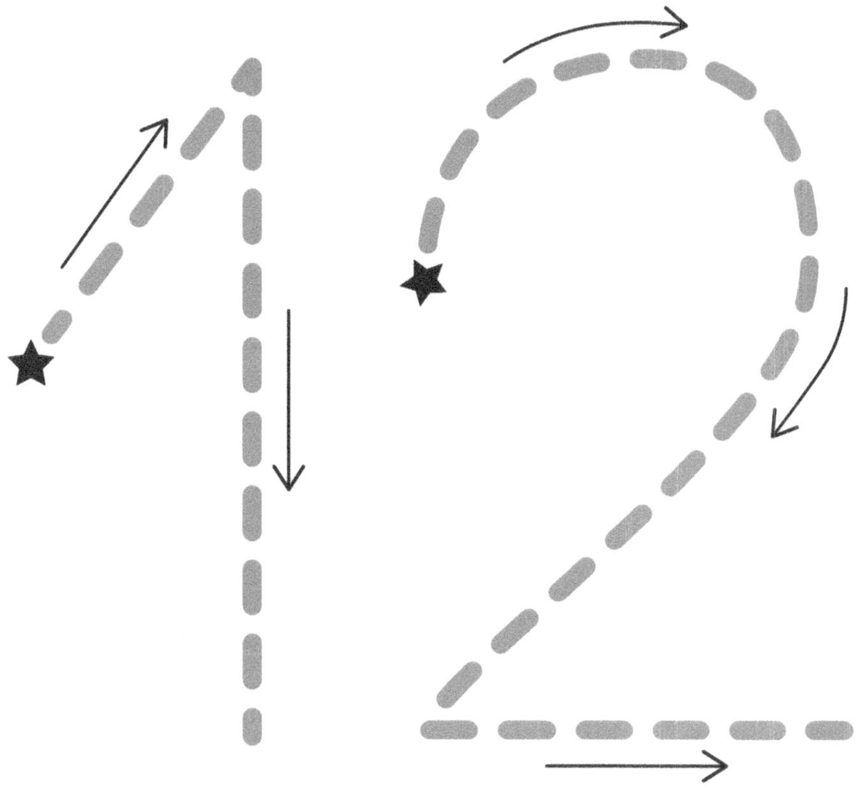
12

12 12 12 12

12 12 12 12

12 12 12 12

Keep drawing flowers
until there are 12. Color them.

thirteen

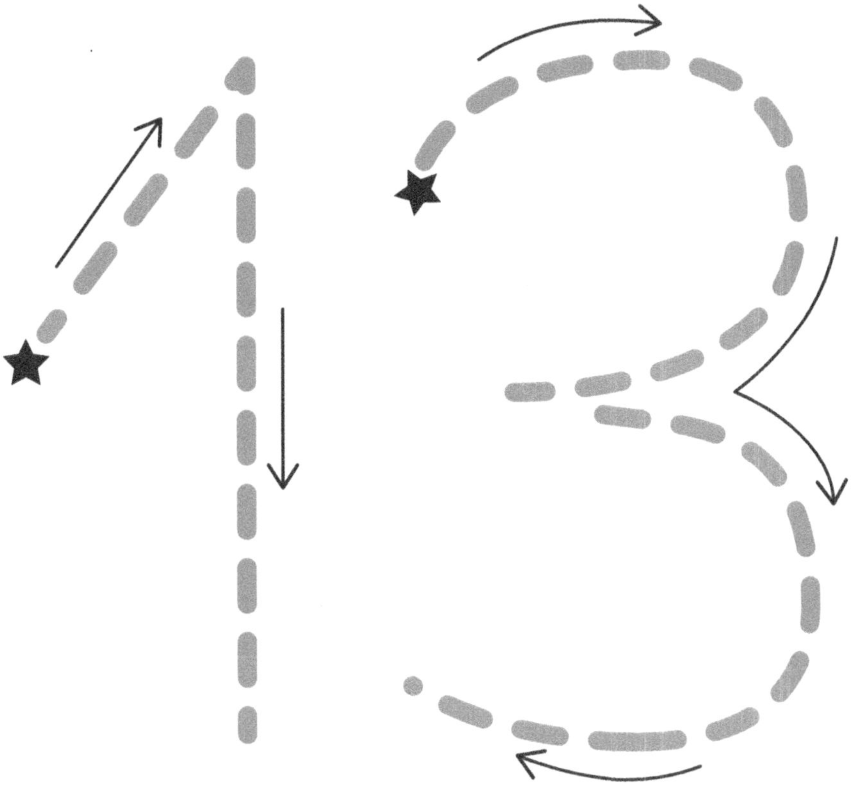

13 13 13 13

13 13 13 13

13 13 13 13

13 13 13 13

13 13 13 13

13 13 13 13

The girl is making soap bubbles.
Trace the 13 bubbles around her.

fourteen

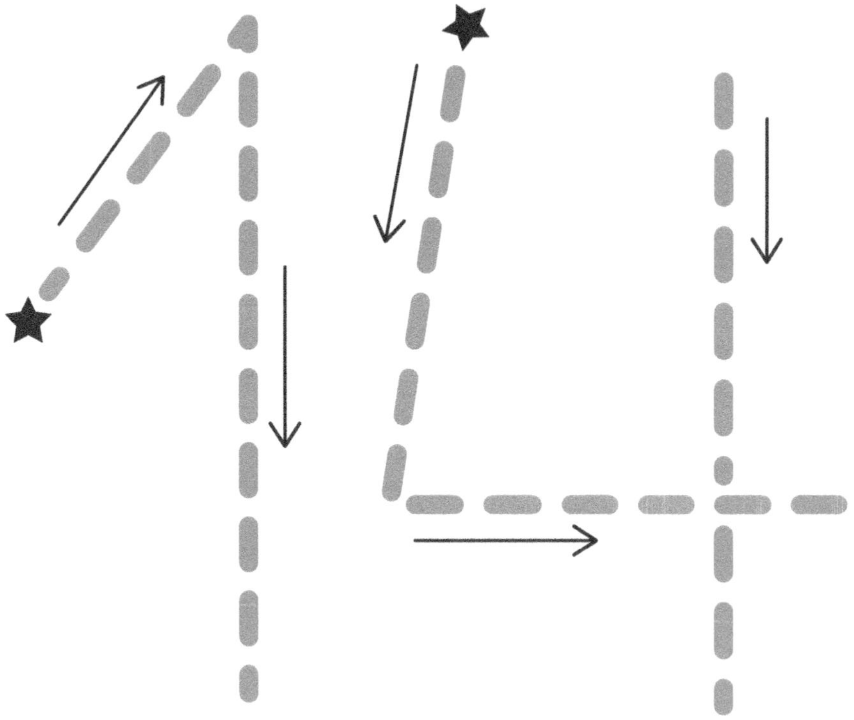

14 14 14 14

14 14 14 14

14 14 14 14

14 14 14 14

Keep drawing leaves
until there are 14 on the tree.

fifteen

15

15 15 15 15

15 15 15 15

15 15 15 15

15 15 15 15

15 15 15 15

15 15 15 15

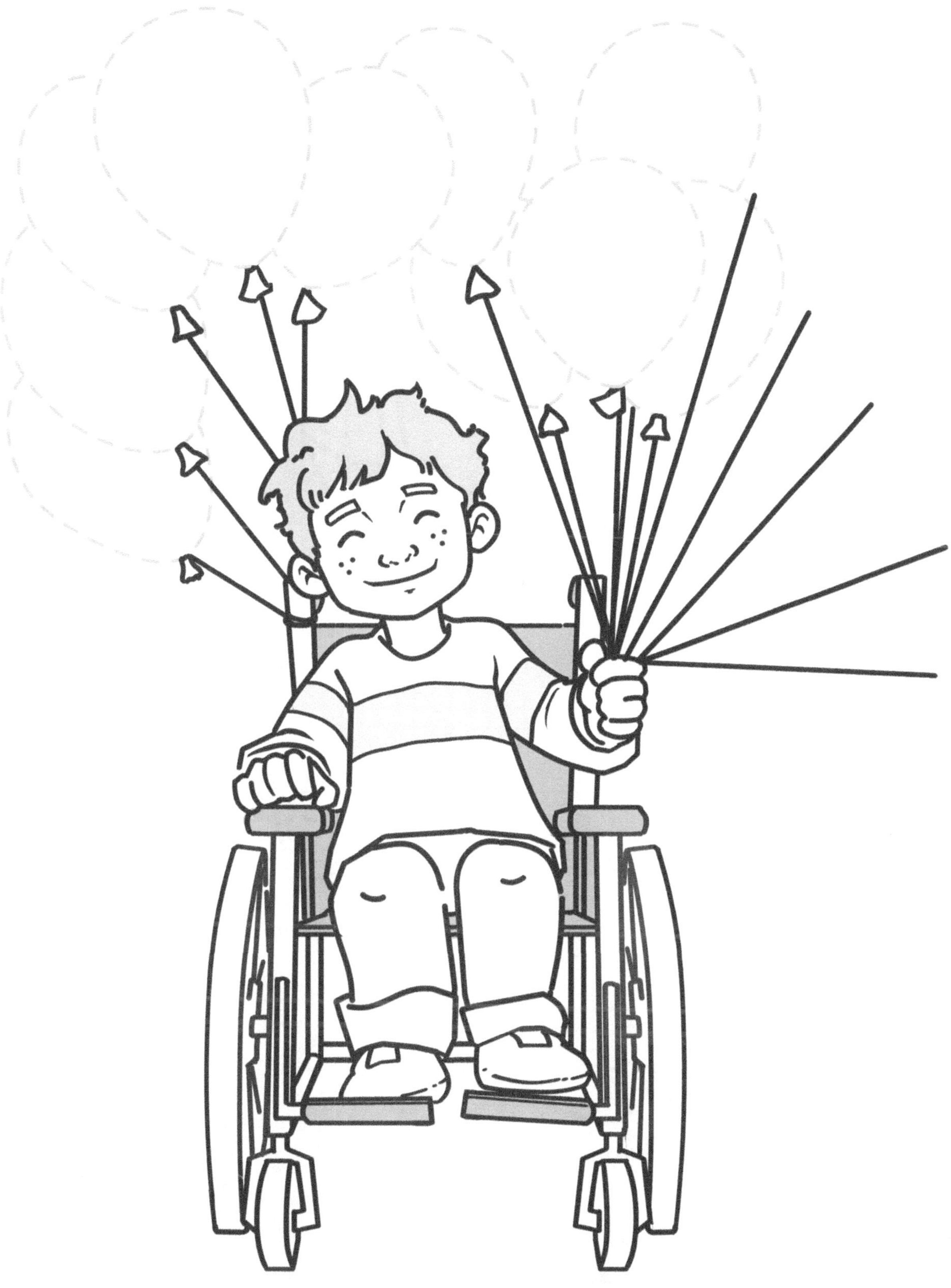

Trace, draw and color

15 balloons.

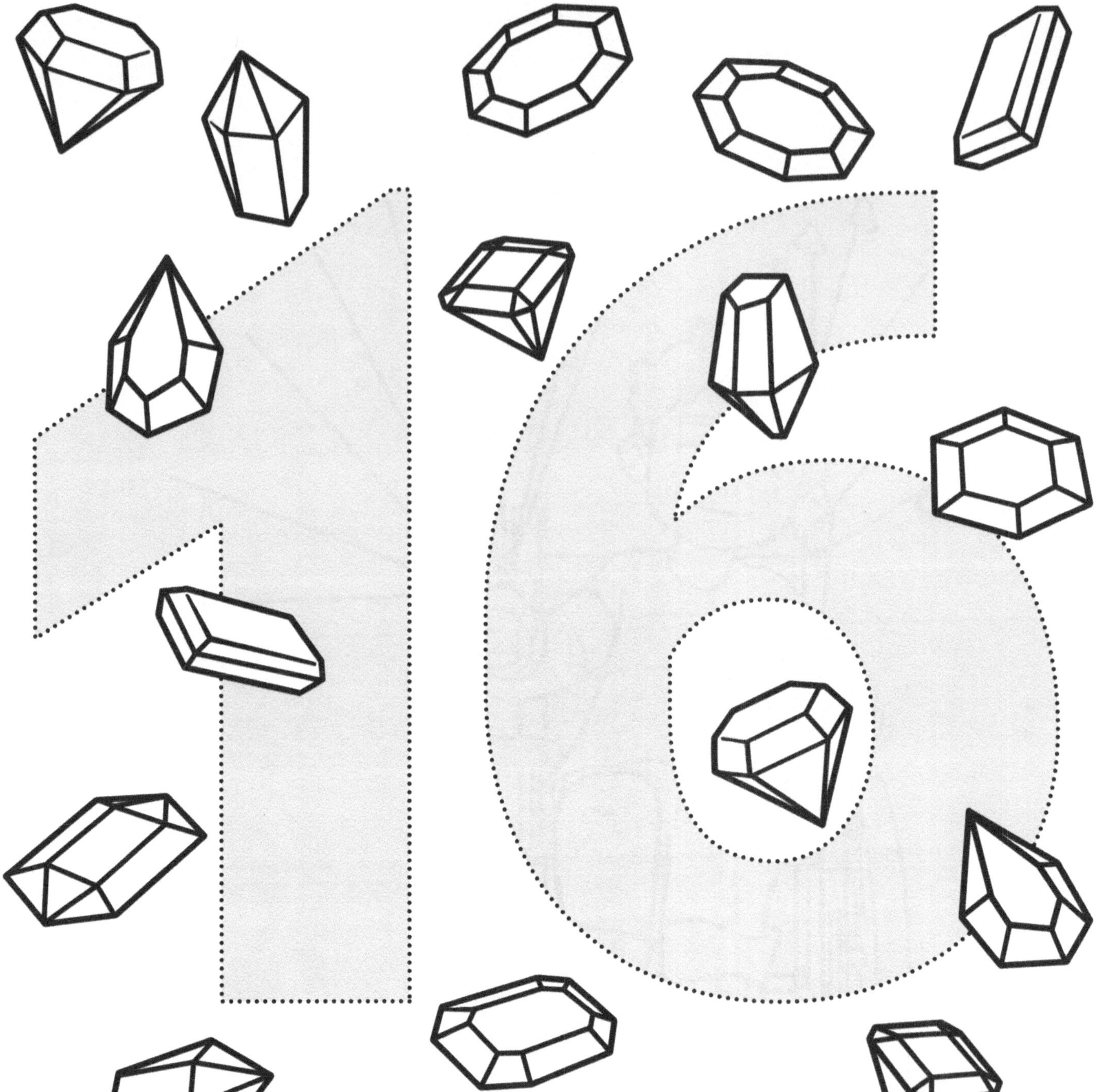

sixteen

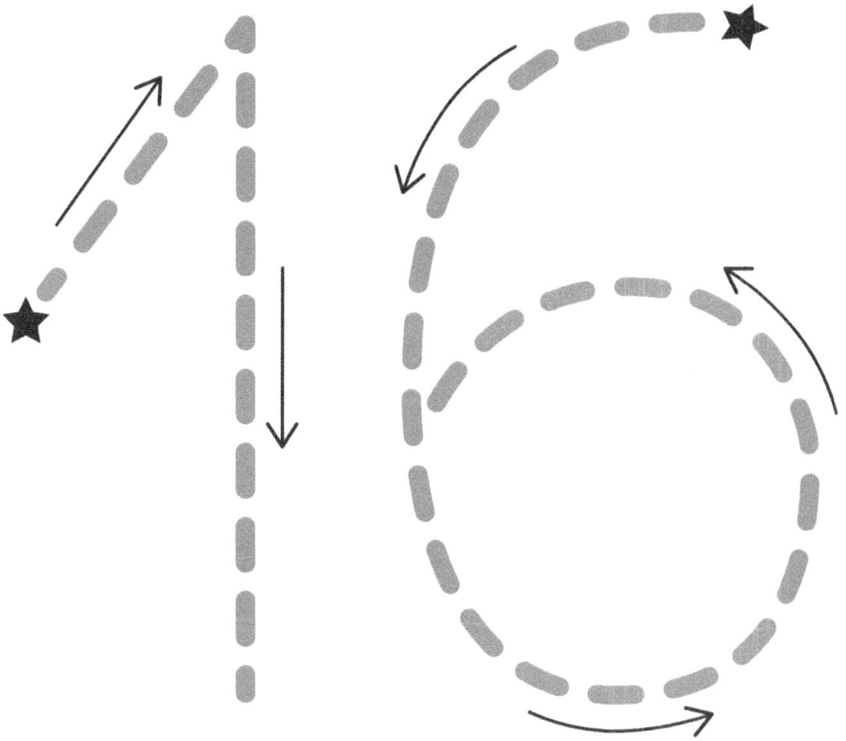

16 16 16 16

16 16 16 16

16 16 16 16

16 16 16 16

Draw 16 coins for the boy's treasure chest.

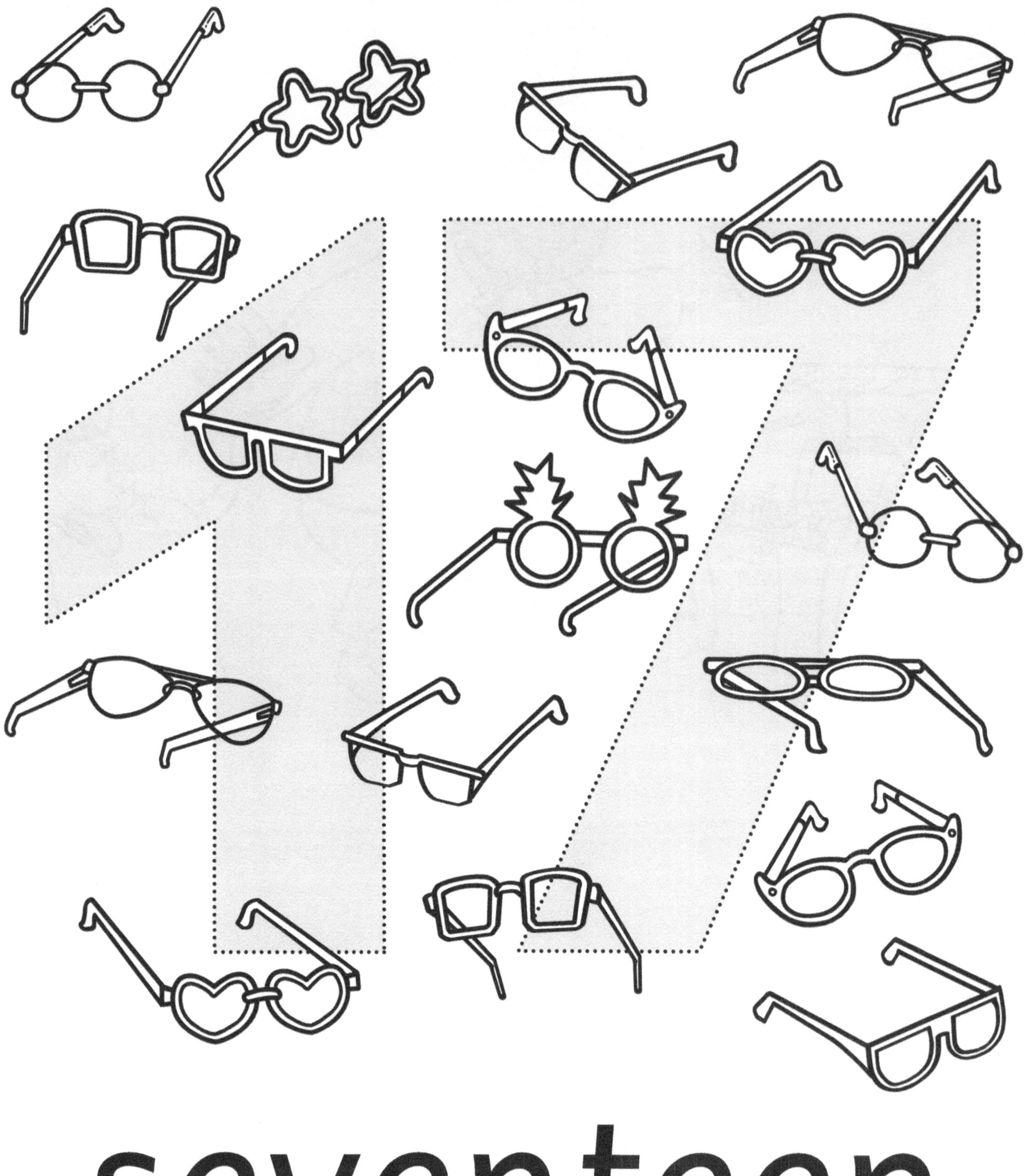

seventeen

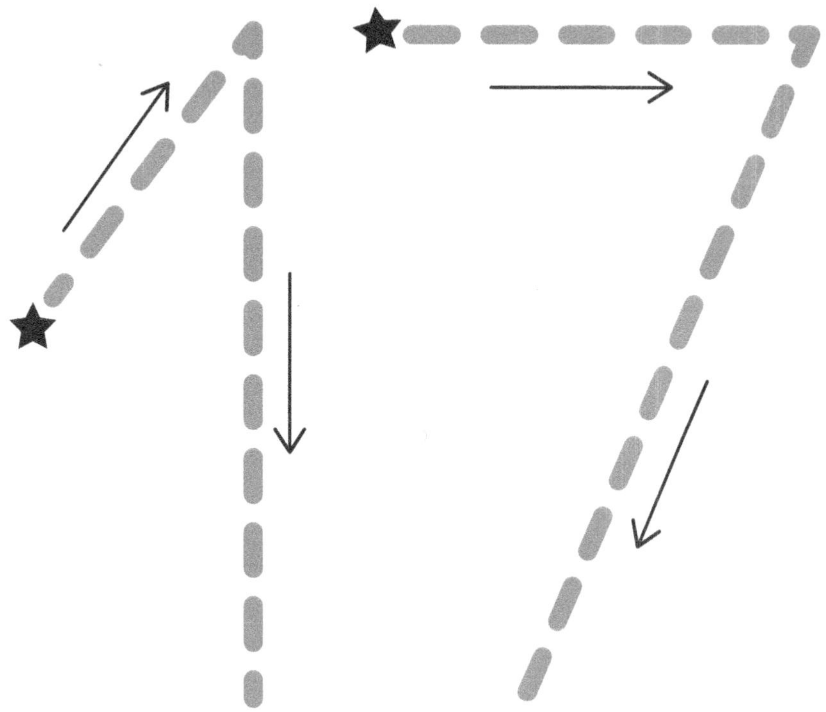
17

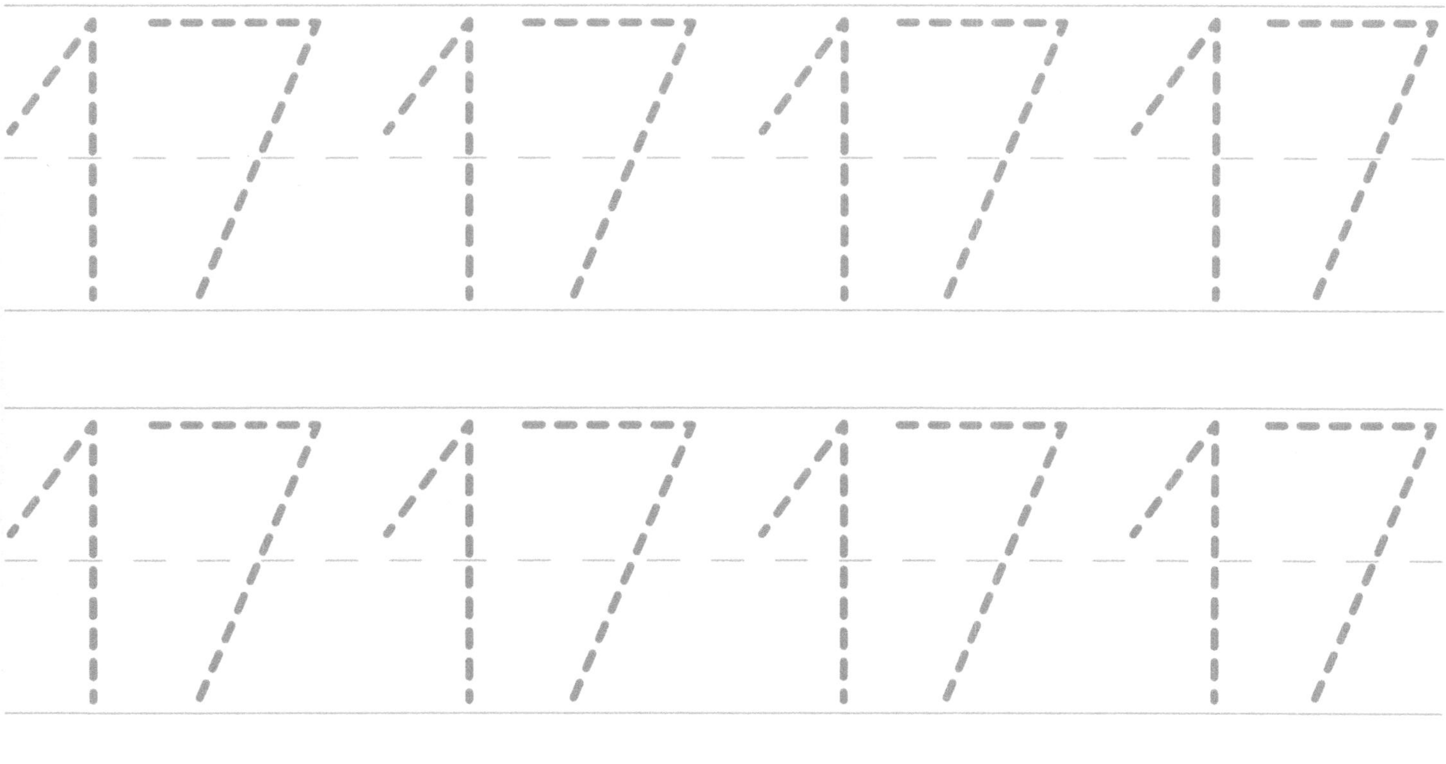
17 17 17 17
17 17 17 17

17 17 17 17

17 17 17 17

17 17 17 17

17 17 17 17

Color all 17 books that the girl is holding.

eighteen

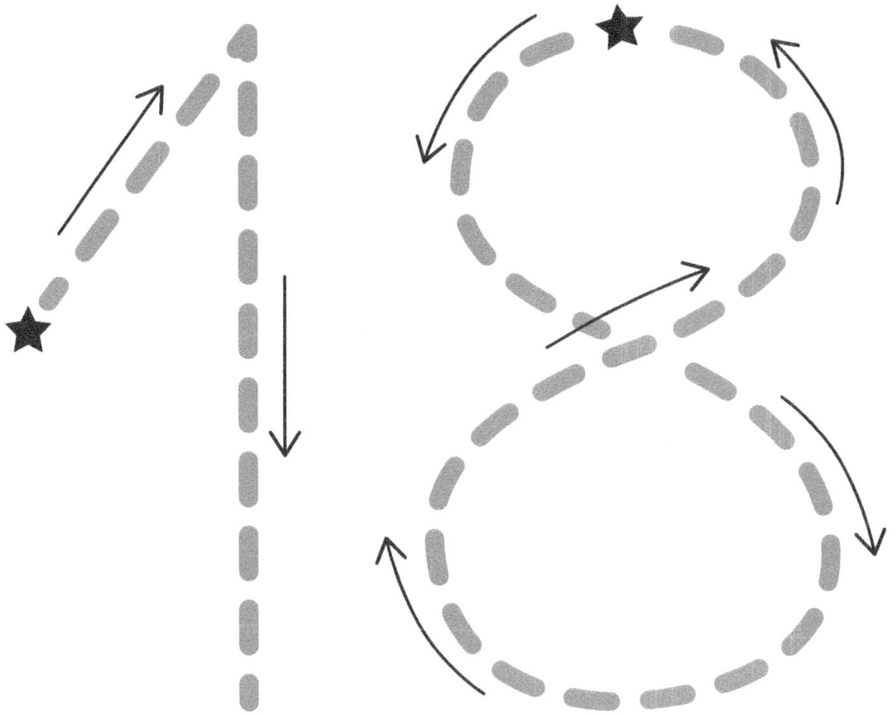
18

18 18 18 18

18 18 18 18

18 18 18 18

18 18 18 18

Keep drawing snowflakes until there are 18 in total.

nineteen

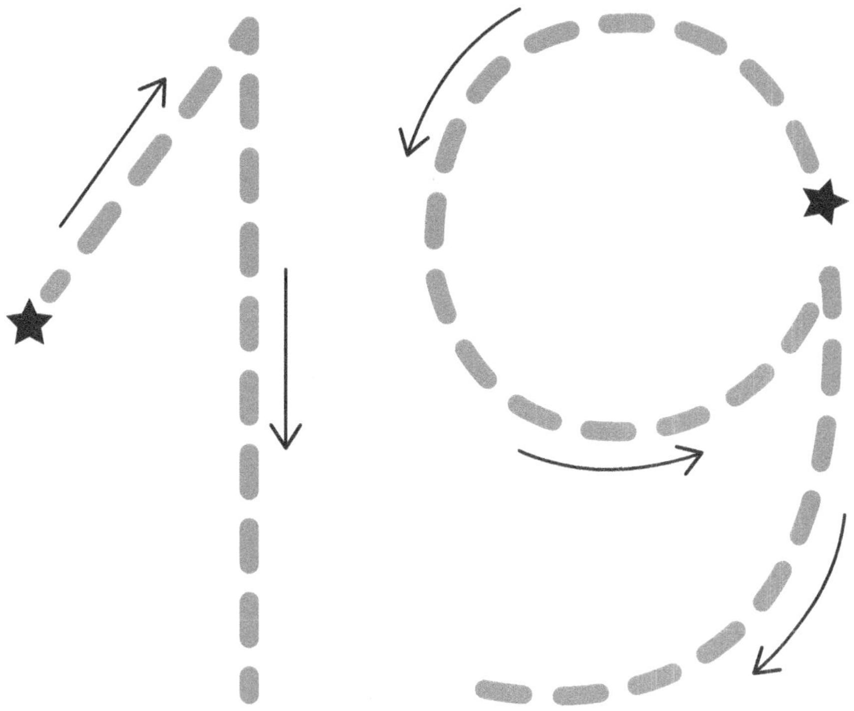

19 19 19 19

19 19 19 19

19 19 19 19

19 19 19 19

Draw 19 rain drops.

twenty

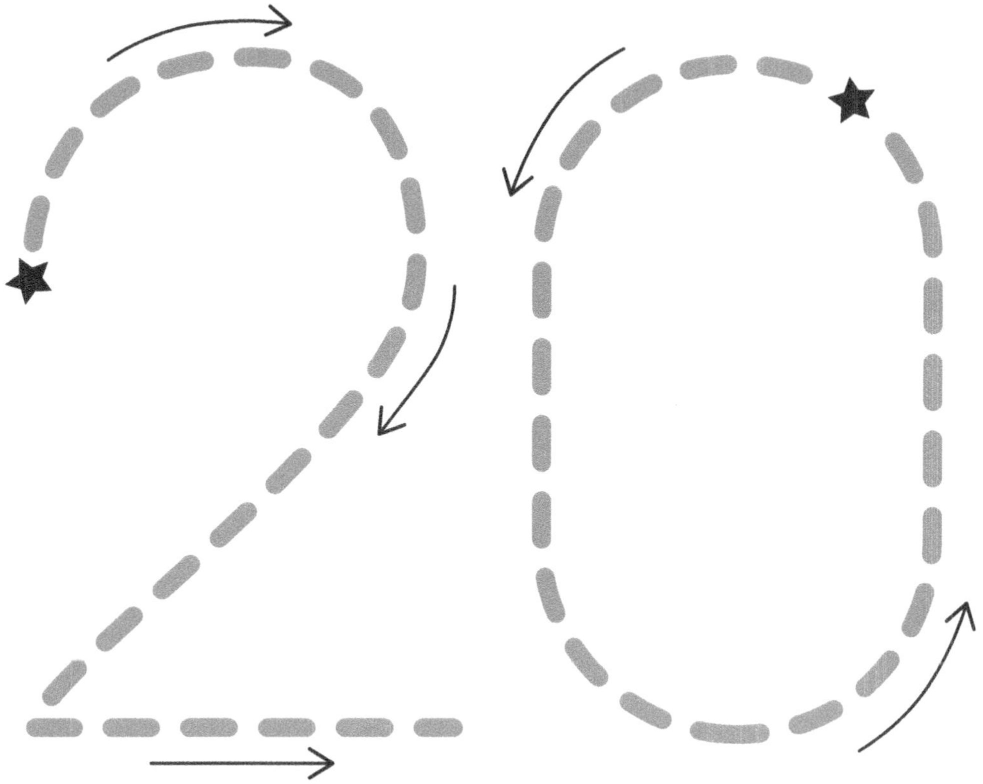
20

20202020

20202020

20202020

20202020

Find and color all 20 kitchen utensils in the picture.

Count and match!

13

15

16

18

19

Fill in the missing numbers.

Count and write!

Count to **100** and trace!

1 2 3 4 5

6 7 8 9 10

11 12 13 14

15 16 17 18

19 20 21 22

23 24 25 26

27 28 29 30

31 32 33 34

35 36 37 38

39 40 41 42

43 44 45 46

47 48 49 50

51 52 53 54

55 56 57 58

59 60 61 62

63 64 65 66

67 68 69 70

71 72 73 74

75 76 77 78

79 80 81 82

83 84 85 86

87 88 89 90

91 92 93 94

95 96 97 98

99 100

Do you know how tall you are?
Write it down.

M	T	W	T	F	S	S
1	2	3	4	5	6	7
8	9	10	11	12	13	14
15	16	17	18	19	20	21
22	23	24	25	26	27	28
29	30	31				

Do you know which day is your birthday?
Circle it on the calendar.
(Feel free to also mark the birthdays of your friends and family!).

A SPECIAL MESSAGE FROM OUR TEAM:

Thank you for bringing our books into your home and hearts!

Your support means the world to us, and fuels our commitment to create educational books of the highest quality for your little ones.

Did you enjoy our book?

If so, we'd be incredibly thankful if you took a moment to leave a review.

With less than 1% of our customers sharing their experiences, your insights are invaluable. Your unique perspective not only helps other parents make informed choices but also guides us on our mission to constantly improve.

Check out other books of our Preschool Learning Activities series:

View details

Made in United States
Cleveland, OH
04 November 2026

47647262R00063